MECHANICS

THINGS YOU SHOULD KNOW
(QUESTIONS AND ANSWERS)

By Rumi Michael Leigh

I would like to thank you for purchasing this book, *"Mechanics, things you should know (questions and answers)".*

This book will help you understand, revise, and have a good general knowledge and understanding of the basics of mechanics.

I hope you enjoy it!

Table of Contents

Part 1: Mechanics

Exercise 1

Questions

a) A 2 kg object is pushed across a rough horizontal surface with a force of 8 N. The coefficient of kinetic friction between the object and the surface is 0.3. What is the acceleration of the object?

b) A ball is thrown horizontally with a velocity of 20 m/s from a height of 10 m. How far from the starting point does it land?

c) A block of mass 3 kg is pushed up a frictionless incline at an angle of 30 degrees with a force of 20 N. What is the acceleration of the block up the incline?

d) A ball is thrown upwards with an initial velocity of 15 m/s from a height of 5 m. How long does it take to hit the ground?

e) A car traveling at a speed of 20 m/s applies its brakes and comes to a stop in a distance of 40 m. What is the magnitude of the car's acceleration?

Answers

a) A 2 kg object is pushed across a rough horizontal surface with a force of 8 N. The coefficient of kinetic friction between the object and the surface is 0.3. What is the acceleration of the object?

The force of friction acting on the object is $F\_friction = \mu\_k * F\_norm$

F_norm is the normal force

The normal force is equal in magnitude to the force of gravity

$F\_gravity = m * g$

g is the acceleration due to gravity.

$F\_norm = F\_gravity = m * g = 2 \text{ kg} * 9.8 \text{ m/s}^2 = 19.6 \text{ N}$.

The force of friction is $F\_friction = 0.3 * 19.6 \text{ N} = 5.88 \text{ N}$.

The net force on the object is $F\_net = 8 \text{ N} - 5.88 \text{ N} = 2.12 \text{ N}$.

Answer: the acceleration of the object is $a = F\_net / m = 2.12 \text{ N} / 2 \text{ kg} = 1.06 \text{ m/s}^2$.

b) A ball is thrown horizontally with a velocity of 20 m/s from a height of 10 m. How far from the starting point does it land?

The horizontal velocity of the ball is constant throughout its motion.

The time it takes for the ball to hit the ground can be found using the equation: $h = 1/2 * g * t^2$, where h is the initial height and g is the acceleration due to gravity.

Solving for t, we get $t = sqrt(2h/g) = sqrt(2*10m/9.8m/s^2) = 1.43$ s.

The distance the ball travels horizontally can be found using the equation: $d = v\_x * t$, where v_x is the horizontal velocity of the ball.

Answer: d = 20 m/s * 1.43 s = 28.6 m.

c) A block of mass 3 kg is pushed up a frictionless incline at an angle of 30 degrees with a force of 20 N. What is the acceleration of the block up the incline?

The force of gravity acting on the block is $F\_gravity = m * g * sin(theta)$

theta is the angle of the incline

g is the acceleration due to gravity.

$F\_gravity = 3$ kg * 9.8 m/s^2 * sin(30 degrees) = 14.7 N.

The force applied to the block up the incline is $F\_applied = 20$ N.

The net force on the block up the incline is :

$F\_net = F\_applied - F\_gravity = 20$ N - 14.7 N = 5.3 N.

The acceleration of the block up the incline is a = F_net / m = 5.3 N / 3 kg = 1.77 m/s^2.

d) A ball is thrown upwards with an initial velocity of 15 m/s from a height of 5 m. How long does it take to hit the ground?

The initial velocity of the ball is $v\_i = 15$ m/s.

The final velocity at impact is $v\_f = 0$ m/s.

The acceleration due to gravity is g = 9.8 m/s^2.

The time it takes for the ball to hit the ground can be found using the equation: $h = v\_i * t - 1/2 * g * t^2$, where h is the initial height.

Solving for t using the quadratic formula:

Answer: $t = (v\_i sqrt(v\_i^2 + 2 * g * h)) / g = (15$ m/s + sqrt(15^2 + 2 * 9.8 m/s^2 * 5 m)) / 9.8 m/s^2 = 2.38 s.

e) A car traveling at a speed of 20 m/s applies its brakes and comes to a stop in a distance of 40 m. What is the magnitude of the car's acceleration?

The car's initial velocity is $v\_i = 20$ m/s.

Its final velocity is $v_f = 0$ m/s.

The distance traveled during deceleration is d = 40 m.

The acceleration of the car can be found using the equation:

$v\_f^2 = v\_i^2 + 2 * a * d$

Answer: a = $(v\_f^2 - v\_i^2) / (2 * d)$ = (0 m/s^2 - 20 m/s^2) / (2 * 40 m) = -0.25 m/s^2

Exercise 2

Questions

a) A block of mass 2 kg is attached to a spring with a spring constant of 50 N/m. The block is pulled 10 cm to the right and released. What is the maximum speed of the block?

b) A ball of mass 0.1 kg is dropped from a height of 2 m onto a spring with a spring constant of 200 N/m. What is the maximum compression of the spring?

c) A projectile is launched with an initial velocity of 30 m/s at an angle of 45 degrees above the horizontal. What is the maximum height reached by the projectile?

d) A crate of mass 50 kg is being pulled up a ramp that is inclined at an angle of 30 degrees to the horizontal. The coefficient of kinetic friction between the crate and the ramp is 0.2. If the pulling force is 500 N, what is the crate's acceleration up the ramp?

e) A ball of mass 0.2 kg is attached to a string and is swung in a vertical circle. If the string is 1 m long, what is the minimum speed required for the ball to complete the circle without falling off?

Answers

a) A block of mass 2 kg is attached to a spring with a spring constant of 50 N/m. The block is pulled 10 cm to the right and released. What is the maximum speed of the block?

The maximum speed of the block occurs when all of the potential energy stored in the spring is converted to kinetic energy.

The potential energy stored in the spring is given by the equation: U_spring = 1/2 * k * x^2, where k is the spring constant and x is the displacement from equilibrium.

U_spring = 1/2 * 50 N/m * (0.1 m)^2 = 0.25 J.

The kinetic energy of the block at the maximum speed is equal to the potential energy stored in the spring.

1/2 * m * v_max^2 = U_spring, where m is the mass of the block.

Answer: v_max, we get v_max = sqrt(2 * U_spring / m) = sqrt(2 * 0.25 J / 2 kg) = 0.5 m/s.

b) A ball of mass 0.1 kg is dropped from a height of 2 m onto a spring with a spring constant of 200 N/m. What is the maximum compression of the spring?

The potential energy of the ball at the initial height is given by the equation: U_initial = m * g * h, where h is the initial height and g is the acceleration due to gravity.

U_initial = 0.1 kg * 9.8 m/s^2 * 2 m = 1.96 J.

The potential energy of the ball is converted to potential energy stored in the spring when the ball hits the spring.

U_spring = 1/2 * k * x^2, where x is the maximum compression of the spring.

Answer: x = sqrt(2 * U_spring / k) = sqrt(2 * 1.96 J / 200 N/m) = 0.14 m.

c) A projectile is launched with an initial velocity of 30 m/s at an angle of 45 degrees above the horizontal. What is the maximum height reached by the projectile?

The initial vertical velocity of the projectile is given by the equation: v_yi = v_i * sin(theta), where v_i is the initial velocity and theta is the launch angle.

Answer: v_yi = 30 m/s * sin(45 degrees) = 21.

d) A crate of mass 50 kg is being pulled up a ramp that is inclined at an angle of 30 degrees to the horizontal. The coefficient of kinetic friction between the crate and the ramp is 0.2. If the pulling force is 500 N, what is the crate's acceleration up the ramp?

The component of the gravitational force acting parallel to the ramp is given by: F_parallel = m * g * sin(theta), where theta is the angle of inclination of the ramp.

F_parallel = 50 kg * 9.8 m/s^2 * sin(30 degrees) = 245 N.

The force of kinetic friction acting on the crate is given by: $F\_friction$ = coefficient of kinetic friction * F_norm, where F_norm is the normal force acting on the crate.

The normal force is equal to the component of the gravitational force perpendicular to the ramp: F_norm = m * g * cos(theta) = 50 kg * 9.8 m/s^2 * cos(30 degrees) = 424 N.

$F\_friction$ = 0.2 * 424 N = 85 N.

The net force acting on the crate is the difference between the pulling force and the force of kinetic friction: F_net = 500 N - 85 N = 415 N.

Answer: the acceleration of the crate up the ramp is given by:

a = F_net / m = 415 N / 50 kg = 8.3 m/s^2.

e) A ball of mass 0.2 kg is attached to a string and is swung in a vertical circle. If the string is 1 m long, what is the minimum speed required for the ball to complete the circle without falling off?

At the top of the circle, the tension in the string must be equal to the weight of the ball: T = m * g, where T is the tension, m is the mass of the ball, and g is the acceleration due to gravity.

T = 0.2 kg * 9.8 m/s^2 = 1.96 N.

At the bottom of the circle, the tension in the string must be greater than the weight of the ball in order to keep the ball moving in a circle: T > m * g.

The centripetal force required to keep the ball moving in a circle is given by: F_c = m * v^2 / r, where v is the speed of the ball and r is the radius of the circle.

T - m * g = m * v^2 / r.

1.96 N - 0.2 kg * 9.8 m/s^2 = 0.2 kg * v^2 / 1 m.

v = sqrt((1.96 N - 1.96 N) / (0.2 kg / 1 m)) = sqrt(0) = 0 m/s.

Answer: the minimum speed required for the ball to complete the circle without falling off is 0 m/s, which means that the ball must be at rest at the top of the circle.

Exercise 3

Questions

a) A car of mass 1000 kg is traveling at 20 m/s. If the brakes are applied and the car comes to a stop in 5 seconds, what is the magnitude of the average braking force?

b) A block of mass 10 kg is pulled along a rough surface by a force of 50 N. The coefficient of kinetic friction between the block and the surface is 0.3. What is the acceleration of the block?

c) A projectile is fired from the ground with an initial velocity of 100 m/s at an angle of 45 degrees to the horizontal. What is the maximum height reached by the projectile?

d) A block of mass 5 kg is pushed up a frictionless incline with an angle of 30 degrees to the horizontal. If the applied force is 50 N, what is the acceleration of the block?

Answer

a) A car of mass 1000 kg is traveling at 20 m/s. If the brakes are applied and the car comes to a stop in 5 seconds, what is the magnitude of the average braking force?

The initial velocity of the car is 20 m/s, the final velocity is 0 m/s, and the time taken to stop is 5 seconds.

The average acceleration of the car is a = (0 m/s - 20 m/s) / 5 s = -4 m/s^2. Answer: using Newton's second law, F = m * a, we can calculate the average braking force as: F = 1000 kg * (-4 m/s^2) = -4000 N.

The negative sign indicates that the force is acting opposite to the direction of motion.

b) A block of mass 10 kg is pulled along a rough surface by a force of 50 N. The coefficient of kinetic friction between the block and the surface is 0.3. What is the acceleration of the block?

The force of friction acting on the block is given by F_friction = coefficient of kinetic friction * F_norm, where F_norm is the normal force acting on the block. The normal force is equal to the weight of the block, which is m * g = 10 kg * 9.8 m/s^2 = 98 N.

F_friction = 0.3 * 98 N = 29.4 N.

The net force acting on the block is F_net = 50 N - 29.4 N = 20.6 N.

The acceleration of the block is given by a = F_net / m = 20.6 N / 10 kg = 2.06 m/s^2.

c) A projectile is fired from the ground with an initial velocity of 100 m/s at an angle of 45 degrees to the horizontal. What is the maximum height reached by the projectile?

The vertical component of the initial velocity is given by V_y = V_0 * sin(theta), where V_0 is the initial velocity and theta is the angle of inclination.

V_y = 100 m/s * sin(45 degrees) = 70.7 m/s.

The time taken for the projectile to reach the maximum height is given by t = V_y / g, where g is the acceleration due to gravity.

t = 70.7 m/s / 9.8 m/s^2 = 7.2 s

The maximum height reached by the projectile is given by h = V_y * t - 0.5 * g * t^2, where t is the time taken to reach the maximum height.

Answer: h = 70.7 m/s * 7.2 s - 0.5 * 9.8 m/s^2 * (7.2 s)^2 = 254.7 m.

d) A block of mass 5 kg is pushed up a frictionless incline with an angle of 30 degrees to the horizontal. If the applied force is 50 N, what is the acceleration of the block?

The weight of the block is given by m * g = 5 kg * 9.8 m/s^2 = 49 N.

The component of the weight acting along the incline is given by F_g = m * g * sin(theta) = 5 kg * 9.8 m/s^2 * sin(30 degrees) = 24.5 N.

The applied force is F_applied = 50 N.

The net force acting on the block is F_net = F_applied - F_g = 50 N - 24.5 N = 25.5 N.

Answer: the acceleration of the block is given by a = F_net / m = 25.5 N / 5 kg = 5.1 m/s^2.

Exercise 4

Questions

a) A car is traveling at a constant speed of 20 m/s around a circular track with a radius of 50 m. What is the magnitude of the centripetal force acting on the car?

b) A mass of 2 kg is attached to a spring with a spring constant of 100 N/m. If the spring is stretched by 0.2 m, what is the magnitude of the force acting on the mass?

c) A block of mass 10 kg is pushed with a force of 50 N across a rough surface. The coefficient of kinetic friction between the block and the surface is 0.2. What is the acceleration of the block?

d) A ball is thrown horizontally off a cliff with a speed of 20 m/s. If the height of the cliff is 100 m, how far away from the base of the cliff will the ball land?

e) A block of mass 5 kg is hanging from a rope. If the tension in the rope is 50 N, what is the acceleration of the block?

Answers

a) A car is traveling at a constant speed of 20 m/s around a circular track with a radius of 50 m. What is the magnitude of the centripetal force acting on the car? The formula for centripetal force is $F\_c = m * v^2 / r$, where m is the mass of the car, v is the speed of the car, and r is the radius of the circular track.
Answer: $F\_c = 1000$ kg $* (20$ m/s$)^2 / 50$ m $= 8000$ N.

b) A mass of 2 kg is attached to a spring with a spring constant of 100 N/m. If the spring is stretched by 0.2 m, what is the magnitude of the force acting on the mass?
The force exerted by the spring is given by Hooke's Law, which states that $F\_spring = -k * x$, where k is the spring constant and x is the displacement of the spring from its equilibrium position.
Answer: $F\_spring = -100$ N/m $* 0.2$ m $= -20$ N.
The negative sign indicates that the force is acting in the opposite direction to the displacement of the spring.

c) A block of mass 10 kg is pushed with a force of 50 N across a rough surface. The coefficient of kinetic friction between the block and the surface is 0.2. What is the acceleration of the block?
The force of friction acting on the block is given by $F\_friction$ = coefficient of kinetic friction $* F\_norm$, where F_norm is the normal force acting on the block. The normal force is equal to the weight of the block, which is $m * g = 10$ kg $* 9.8$ m/s$^2 = 98$ N. Therefore, $F\_friction = 0.2 * 98$ N $= 19.6$ N.

The net force acting on the block is F_net = 50 N - 19.6 N = 30.4 N.

The acceleration of the block is given by a = F_net / m = 30.4 N / 10 kg = 3.04 m/s^2.

d) A ball is thrown horizontally off a cliff with a speed of 20 m/s. If the height of the cliff is 100 m, how far away from the base of the cliff will the ball land?

The time taken for the ball to hit the ground is given by t = sqrt(2h / g), where h is the height of the cliff and g is the acceleration due to gravity.

t = sqrt(2 * 100 m / 9.8 m/s^2) = 4.52 s.

The horizontal distance traveled by the ball is given by d = v * t, where v is the horizontal velocity of the ball, which is 20 m/s.

Answer: d = 20 m/s * 4.52 s = 90.4 m

e) A block of mass 5 kg is hanging from a rope. If the tension in the rope is 50 N, what is the acceleration of the block?

The weight of the block is given by m * g = 5 kg * 9.8 m/s^2 = 49 N.

The net force acting on the block is F_net = T - m * g, where T is the tension in the rope.

F_net = 50 N - 49 N = 1 N.

Answer: the acceleration of the block is = F_net / m = 1 N / 5 kg = 0.2 m/s^2.

Exercise 5

Questions

a) A car is traveling at a speed of 30 m/s when the driver slams on the brakes. If the coefficient of kinetic friction between the tires and the road is 0.5, how far will the car travel before coming to a stop?

b) A block of mass 2 kg is pushed up a rough incline with an angle of 30 degrees to the horizontal. The coefficient of kinetic friction between the block and the incline is 0.2. If the block is pushed with a force of 20 N, what is the acceleration of the block?

c) A ball is thrown vertically upwards with an initial velocity of 20 m/s. How high does the ball rise before it begins to fall back down?

d) A block of mass 10 kg is pushed with a force of 50 N along a frictionless surface. What is the acceleration of the block?

e) A car is traveling at a speed of 60 km/h. What is its speed in meters per second?

60 km/h is equivalent to 16.67 m/s.

Answers

a) A car is traveling at a speed of 30 m/s when the driver slams on the brakes. If the coefficient of kinetic friction between the tires and the road is 0.5, how far will the car travel before coming to a stop?

The deceleration of the car is given by a = -coefficient of kinetic friction * g = - 0.5 * 9.8 m/s^2 = -4.9 m/s^2.

The time taken for the car to come to a stop is given by t = v / a, where v is the initial velocity of the car.

t = 30 m/s / 4.9 m/s^2 = 6.12 s

The distance traveled by the car is given by d = v * t + 1/2 * a * t^2

Answer: d = 30 m/s * 6.12 s + 1/2 * (-4.9 m/s^2) * (6.12 s)^2 = 92.4 m

b) A block of mass 2 kg is pushed up a rough incline with an angle of 30 degrees to the horizontal. The coefficient of kinetic friction between the block and the incline is 0.2. If the block is pushed with a force of 20 N, what is the acceleration of the block?

The component of the weight of the block acting along the incline is given by F_g = m * g * sin(theta) = 2 kg * 9.8 m/s^2 * sin(30 degrees) = 9.8 N.

The force of friction acting on the block is given by F_friction = coefficient of kinetic friction * F_norm, where F_norm is the normal force acting on the block. The normal force is equal to the weight of the block, which is m * g = 2 kg * 9.8 m/s^2 = 19.6 N.

F_friction = 0.2 * 19.6 N = 3.92 N.

The net force acting on the block is F_net = F_applied - F_g - F_friction = 20 N - 9.8 N - 3.92 N = 6.28 N.

Answer: the acceleration of the block is given by a = F_net / m = 6.28 N / 2 kg = 3.14

c) A ball is thrown vertically upwards with an initial velocity of 20 m/s. How high does the ball rise before it begins to fall back down?

$v^2 = u^2 + 2as$, where u is the initial velocity, v is the final velocity (which is 0 when the ball reaches its highest point), a is the acceleration due to gravity (-9.8 m/s^2), and s is the displacement, we can solve for s.

s = (0 - 20^2) / (2 * -9.8) = 20.4 meters.

Answer: the ball rises to a height of 20.4 meters before it begins to fall back down.

d) A block of mass 10 kg is pushed with a force of 50 N along a frictionless surface. What is the acceleration of the block?

The acceleration of the block is given by a = F_net / m, where F_net is the net force acting on the block and m is its mass.

In this case, since there is no frictional force, the net force is equal to the applied force of 50 N.

Answer: a = 50 N / 10 kg = 5 m/s^2

e) A car is traveling at a speed of 60 km/h. What is its speed in meters per second?

60 km/h is equivalent to 16.67 m/s.

Convert from km/h to m/s

60 km/h / 3.6 = 16.67 m/s.

Answer: 16.67 m/s.

Part 2: Mechanics

Exercise 1

Questions

a) A ball is thrown horizontally with a speed of 10 m/s from the top of a cliff that is 50 meters high. How far from the base of the cliff does the ball land?

b) A block of mass 5 kg is pulled along a rough surface with a force of 20 N. The coefficient of kinetic friction between the block and the surface is 0.4. What is the acceleration of the block?

c) A ball is dropped from a height of 40 meters. What is its speed just before it hits the ground?

d) A car accelerates from rest to a speed of 20 m/s in 5 seconds. What is its acceleration?

e) A block of mass 2 kg is placed on an inclined plane that makes an angle of 30 degrees with the horizontal. If the coefficient of static friction between the block and the plane is 0.5, what is the maximum angle of inclination for which the block will not slide down the plane?

Answers

a) A ball is thrown horizontally with a speed of 10 m/s from the top of a cliff that is 50 meters high. How far from the base of the cliff does the ball land?

Since the ball is thrown horizontally, its initial vertical velocity is 0.

t = sqrt(2h / g), where h is the height of the cliff and g is the acceleration due to gravity.

t = sqrt(2 * 50 m / 9.8 m/s^2) = 3.19 s

The horizontal distance traveled by the ball is given by d = v * t, where v is the horizontal velocity of the ball, which is 10 m/s.

d = 10 m/s * 3.19 s = 31.9 meters

Answer: 31.9 meters

b) A block of mass 5 kg is pulled along a rough surface with a force of 20 N. The coefficient of kinetic friction between the block and the surface is 0.4. What is the acceleration of the block?

The force of friction acting on the block is given by F_friction = coefficient of kinetic friction * F_norm, where F_norm is the normal force acting on the block. The normal force is equal to the weight of the block, which is m * g = 5 kg * 9.8 m/s^2 = 49 N.

F_friction = 0.4 * 49 N = 19.6 N.

The net force acting on the block is F_net = F_applied - F_friction = 20 N - 19.6 N = 0.4 N.

Answer: the acceleration of the block is given by a = F_net / m = 0.4 N / 5 kg = 0.08 m/s^2

c) A ball is dropped from a height of 40 meters. What is its speed just before it hits the ground?

The final velocity of the ball just before it hits the ground is given by v = sqrt(2gh), where g is the acceleration due to gravity and h is the height from which the ball is dropped.

Answer: v = sqrt(2 * 9.8 m/s^2 * 40 m) = 28.0 m/s.

d) A car accelerates from rest to a speed of 20 m/s in 5 seconds. What is its acceleration?

The acceleration of the car is given by a = (v_f - v_i) / t, where v_f is the final velocity, v_i is the initial velocity, and t is the time taken to change the velocity.

Answer: a = (20 m/s - 0 m/s) / 5 s = 4 m/s^2.

e) A block of mass 2 kg is placed on an inclined plane that makes an angle of 30 degrees with the horizontal. If the coefficient of static friction between the block and the plane is 0.5, what is the maximum angle of inclination for which the block will not slide down the plane?

The maximum angle of inclination for which the block will not slide down the plane is given by theta = arctan(mu_s), where mu_s is the coefficient of static friction.

theta = arctan(0.5) = 26.6 degrees

Answer: 26.6 degrees

Exercise 2

Questions

a) A ball is thrown horizontally with a speed of 20 m/s from a height of 5 meters above the ground. How far from the base of the cliff does the ball land?

b) A force of 100 N is applied to a block of mass 5 kg. What is its acceleration?

c) A ball is thrown vertically upwards with an initial velocity of 30 m/s. How long does it take to reach its maximum height?

d) A block of mass 4 kg is pulled along a rough surface with a force of magnitude 20 N. If the coefficient of kinetic friction between the block and the surface is 0.3, what is its acceleration?

e) A bullet of mass 10 g is fired into a wooden block of mass 2 kg. If the bullet has an initial velocity of 500 m/s and it gets embedded into the block, what is the velocity of the block just after the collision?

Answers

a) A ball is thrown horizontally with a speed of 20 m/s from a height of 5 meters above the ground. How far from the base of the cliff does the ball land?

The horizontal distance traveled by the ball is given by $d = v\_x * t$, where v_x is the horizontal velocity of the ball and t is the time it takes to hit the ground. Since the ball is thrown horizontally, its initial vertical velocity is 0.

The time it takes to hit the ground is given by $t = sqrt(2h / g)$, where h is the height from which the ball is thrown and g is the acceleration due to gravity.

$t = sqrt(2 * 5 m / 9.8 m/s^2) = 1.01 s$

The horizontal velocity of the ball is 20 m/s.

Answer: $d = 20 m/s * 1.01 s = 20.2$ meters.

b) A force of 100 N is applied to a block of mass 5 kg. What is its acceleration?

The acceleration of the block is given by $a = F\_net / m$, where F_net is the net force acting on the block and m is its mass.

In this case, the net force is equal to the applied force of 100 N.

Answer: $a = 100 N / 5 kg = 20 m/s^2$.

c) A ball is thrown vertically upwards with an initial velocity of 30 m/s. How long does it take to reach its maximum height?

The time taken by the ball to reach its maximum height is given by $t = v / g$, where v is the initial vertical velocity and g is the acceleration due to gravity.

Answer: $t = 30 \text{ m/s} / 9.8 \text{ m/s}^2 = 3.06$ seconds.

d) A block of mass 4 kg is pulled along a rough surface with a force of magnitude 20 N. If the coefficient of kinetic friction between the block and the surface is 0.3, what is its acceleration?

The net force acting on the block is given by $F\_net = F\_applied - F\_friction$, where $F\_applied$ is the applied force and $F\_friction$ is the force of friction.

The force of friction is given by $F\_friction = mu\_k * N$, where mu_k is the coefficient of kinetic friction and N is the normal force.

The normal force is equal to the weight of the block, which is given by $m * g$, where m is the mass of the block and g is the acceleration due to gravity.

$F\_friction = mu\_k * m * g$

$F\_friction = 0.3 * 4 \text{ kg} * 9.8 \text{ m/s}^2 = 11.76 \text{ N}$

The net force is then given by $F\_net = 20 \text{ N} - 11.76 \text{ N} = 8.24 \text{ N}$.

The acceleration of the block is given by $a = F\_net / m$

Answer: $a = 8.24 \text{ N} / 4 \text{ kg} = 2.06 \text{ m/s}^2$.

e) A bullet of mass 10 g is fired into a wooden block of mass 2 kg. If the bullet has an initial velocity of 500 m/s and it gets embedded into the block, what is the velocity of the block just after the collision?

The total momentum of the bullet-block system before the collision is given by $p\_i = m\_bullet * v\_bullet$, where m_bullet is the mass of the bullet and v_bullet is its initial velocity.

After the collision, the bullet gets embedded into the block and the total momentum of the system is given by $p\_f = (m\_bullet + m\_block) * v\_f$, where m_block is the mass of the block and v_f is its final velocity.

By conservation of momentum, we have $p\_i = p\_f$.

$(10 \text{ g}) * (500 \text{ m/s}) = (10 \text{ g} + 2 \text{ kg}) * v\_f$

Answer: $v\_f = 2.5 \text{ m/s}$

Exercise 3

Questions

a) A rocket accelerates from rest at a rate of 10 m/s^2 for 5 seconds. What is its final velocity?

b) A block of mass 5 kg is pushed up an inclined plane that makes an angle of 30 degrees with the horizontal. If the coefficient of kinetic friction between the block and the plane is 0.2 and it takes 10 seconds for the block to travel a distance of 20 meters up the plane, what is the magnitude of the force applied to the block?

c) A car traveling at a speed of 20 m/s is brought to rest by applying brakes. If the deceleration of the car is 5 m/s^2, how far will it travel before coming to rest?

d) A ball is thrown vertically upward with a velocity of 20 m/s. What is the maximum height it will reach?

e) A force of 10 N is applied to a block of mass 2 kg, initially at rest, for 5 seconds. What is the final velocity of the block?

Answers

a) A rocket accelerates from rest at a rate of 10 m/s^2 for 5 seconds. What is its final velocity?

The final velocity of the rocket is given by v_f = v_i + a * t, where v_i is the initial velocity, a is the acceleration, and t is the time taken.

Since the rocket starts from rest, its initial velocity is 0.

Answer: v_f = 0 + 10 m/s^2 * 5 s = 50 m/s.

b) A block of mass 5 kg is pushed up an inclined plane that makes an angle of 30 degrees with the horizontal. If the coefficient of kinetic friction between the block and the plane is 0.2 and it takes 10 seconds for the block to travel a distance of 20 meters up the plane, what is the magnitude of the force applied to the block?

The work done by the applied force on the block is given by W = F_applied * d * cos(theta), where F_applied is the applied force, d is the distance traveled by the block, and theta is the angle between the applied force and the displacement of the block.

Since the block is moving up the plane, the angle between the applied force and the displacement is 30 degrees.

The work done by the force of friction is given by W_friction = F_friction * d * cos(180degrees), where F_friction is the force of friction.

The net work done on the block is given by W_net = W - W_friction.

By the work-energy theorem, we have W_net = (1/2) * m * v_f^2, where m is the mass of the block and v_f is its final velocity.

Since the block is moving at a constant velocity, its final kinetic energy is equal to its initial kinetic energy, which is 0.

Therefore, we have W_net = 0.

Substituting the given values, we get F_applied * 20 m * cos(30 degrees) - 0.2 * m * g * 20 m * cos(180 degrees) = 0

Answer: F_applied = (0.2 * 5 kg * 9.8 m/s^2) / cos(30 degrees) = 43.03 N

c) A car traveling at a speed of 20 m/s is brought to rest by applying brakes. If the deceleration of the car is 5 m/s^2, how far will it travel before coming to rest?

The distance traveled by the car before coming to rest is given by the formula, d = (v_f^2 - v_i^2) / (2 * a), where v_i is the initial velocity, v_f is the final velocity, and a is the acceleration (deceleration in this case).

Here, v_i = 20 m/s, v_f = 0 m/s, and a = -5 m/s^2 (negative because it is a deceleration).

d = (0^2 - 20^2) / (2 * -5) = 40 m. Therefore, the car will travel 40 meters before coming to rest.

d) A ball is thrown vertically upward with a velocity of 20 m/s. What is the maximum height it will reach?

The maximum height reached by the ball can be calculated using the formula, h = (v_i^2) / (2 * g), where v_i is the initial velocity and g is the acceleration due to gravity.

v_i = 20 m/s and g = 9.8 m/s^2

h = (20^2) / (2 * 9.8) = 20.41 meters

Answer: the ball will reach a maximum height of 20.41 meters.

e) A force of 10 N is applied to a block of mass 2 kg, initially at rest, for 5 seconds. What is the final velocity of the block?

The final velocity of the block can be calculated using the formula, v_f = v_i + (F_net / m) * t, where v_i is the initial velocity, F_net is the net force acting on the block, m is the mass of the block, and t is the time taken.

v_i = 0 m/s

F_net = 10 N

m = 2 kg

t = 5 seconds.

v_f = 0 + (10 / 2) * 5 = 25 m/s.

Answer: the final velocity of the block is 25 m/s.

Exercise 4

Questions

a) A car travels 100 km at a constant speed of 50 km/hr. What is the time taken by the car to cover this distance?

b) A bullet is fired from a gun with an initial velocity of 500 m/s. If the bullet hits a target 500 meters away, how long does it take for the bullet to reach the target?

c) A ball is thrown horizontally with a speed of 10 m/s from a height of 20 meters. How far away from the point of projection will the ball hit the ground?

d) A force of 20 N is applied to a block of mass 4 kg, initially at rest, for 2 seconds. What is the change in the block's momentum?

e) A stone is dropped from a height of 100 meters. What is the velocity of the stone just before it hits the ground?

Answers

a) A car travels 100 km at a constant speed of 50 km/hr. What is the time taken by the car to cover this distance?

The time taken by the car can be calculated using the formula, time = distance / speed. Here, distance = 100 km and speed = 50 km/hr.

time = 100 km / 50 km/hr = 2 hours

Answer: the car takes 2 hours to cover a distance of 100 km.

b) A bullet is fired from a gun with an initial velocity of 500 m/s. If the bullet hits a target 500 meters away, how long does it take for the bullet to reach the target?

The time taken by the bullet to reach the target can be calculated using the formula, time = distance / speed.

distance = 500 m and speed = 500 m/s

time = 500 m / 500 m/s = 1 second.

Answer: the bullet takes 1 second to reach the target.

c) A ball is thrown horizontally with a speed of 10 m/s from a height of 20 meters. How far away from the point of projection will the ball hit the ground?

The time taken by the ball to hit the ground can be calculated using the formula, $h = (1/2) * g * t^2$, where h is the initial height, g is the acceleration due to gravity, and t is the time taken.

h = 20 meters and g = 9.8 m/s^2

$t = sqrt((2 * h) / g) = sqrt((2 * 20) / 9.8) = 2.02$ seconds (rounded to 2 decimal places).

The horizontal distance covered by the ball can be calculated using the formula, distance = speed * time, where speed is the horizontal component of the initial velocity.

speed = 10 m/s (since the ball is thrown horizontally) and time = 2.02 seconds.

distance = 10 m/s * 2.02 s = 20.2 meters.

Answer: the ball will hit the ground 20.2 meters away from the point of projection.

d) A force of 20 N is applied to a block of mass 4 kg, initially at rest, for 2 seconds. What is the change in the block's momentum?

The change in momentum of the block can be calculated using the formula, delta p = F_net * t, where F_net is the net force acting on the block and t is the time taken.

F_net = 20 N and t = 2 seconds.

The mass of the block is 4 kg, so its initial momentum is 0.

delta p = 20 N * 2 s = 40 Ns.

Answer: the change in momentum of the block is 40 Ns.

e) A stone is dropped from a height of 100 meters. What is the velocity of the stone just before it hits the ground?

The velocity of the stone just before it hits the ground can be calculated using the formula, $v_f = sqrt(2 * g * h)$, where g is the acceleration due to gravity and h is the initial height.

g = 9.8 m/s^2 and h = 100 meters.

$v_f = sqrt(2 * 9.8 * 100) = 44.3$ m/s.

Answer: the velocity of the stone just before it hits the ground is 44.3 m/s.

Exercise 5

Questions

a) A force of 50 N is applied to a block of mass 10 kg, initially at rest, on a frictionless surface. What is the acceleration of the block?

b) A car is traveling at a speed of 72 km/hr. What is the speed of the car in meters per second?

c) A ball is thrown straight up into the air with a speed of 20 m/s. How high does it go?

d) A car is traveling at a speed of 40 km/hr. What is the speed of the car in meters per second?

e) A block of mass 2 kg is pushed with a force of 10 N. What is its acceleration?

Answers

a) A force of 50 N is applied to a block of mass 10 kg, initially at rest, on a frictionless surface. What is the acceleration of the block?

The acceleration of the block can be calculated using the formula, $a = F\_net / m$, where F_net is the net force acting on the block and m is the mass of the block.

$F\_net = 50$ N and $m = 10$ kg.

$a = 50$ N $/ 10$ kg $= 5$ m/s^2.

Answer: the acceleration of the block is 5 m/s^2.

b) A car is traveling at a speed of 72 km/hr. What is the speed of the car in meters per second?

The speed of the car in meters per second can be calculated by converting the units using the formula, 1 km/hr = 0.2778 m/s.

Answer: the speed of the car in meters per second is 72 km/hr * 0.2778 m/s/km/hr = 20 m/s.

c) A ball is thrown straight up into the air with a speed of 20 m/s. How high does it go?

The height the ball goes can be calculated using the formula, $h = (v\_f^2 - v\_i^2) / (2 * g)$, where v_i is the initial velocity, v_f is the final velocity (which is 0 at the highest point), and g is the acceleration due to gravity.

v_i = 20 m/s and g = 9.8 m/s^2.

h = (0 - 20^2) / (2 * -9.8) = 20.4 meters.

Answer: the ball goes to a height of 20.4 meters.

d) A car is traveling at a speed of 40 km/hr. What is the speed of the car in meters per second?

The speed of the car in meters per second can be calculated by converting the units using the formula, 1 km/hr = 0.2778 m/s.

Answer: the speed of the car in meters per second is 40 km/hr * 0.2778 m/s/km/hr = 11.1 m/s.

e) A block of mass 2 kg is pushed with a force of 10 N. What is its acceleration?

The acceleration of the block can be calculated using the formula, a = F_net / m, where F_net is the net force acting on the block and m is the mass of the block.

F_net = 10 N and m = 2 kg.

a = 10 N / 2 kg = 5 m/s^2.

Answer: the acceleration of the block is 5 m/s^2

Part 3: Mechanics

Exercise 1

Questions

a) A cyclist travels a distance of 15 km in 30 minutes. What is the average speed of the cyclist in meters per second?

b) A ball is dropped from a height of 5 meters. What is its speed just before it hits the ground?

c) A car is moving at a speed of 20 m/s. If the brakes are applied and the car comes to a stop in 5 seconds, what is the deceleration of the car?

d) A stone is dropped from a height of 20 meters. How much time does it take to reach the ground?

e) A box of mass 10 kg is pushed with a force of 50 N. What is the acceleration of the box?

Answers

a) A cyclist travels a distance of 15 km in 30 minutes. What is the average speed of the cyclist in meters per second?

The average speed of the cyclist in meters per second can be calculated by converting the units using the formula, 1 km = 1000 m and 1 min = 60 s.
Answer: the average speed of the cyclist is (15 km * 1000 m/km) / (30 min * 60 s/min) = 8.3 m/s.

b) A ball is dropped from a height of 5 meters. What is its speed just before it hits the ground?

The speed of the ball just before it hits the ground can be calculated using the formula, $v_f = sqrt(2 * g * h)$, where g is the acceleration due to gravity and h is the initial height.

$g = 9.8$ m/s^2 and $h = 5$ meters.

$v_f = sqrt(2 * 9.8 * 5) = 9.9$ m/s.

Answer: the speed of the ball just before it hits the ground is 9.9 m/s.

c) A car is moving at a speed of 20 m/s. If the brakes are applied and the car comes to a stop in 5 seconds, what is the deceleration of the car?

The deceleration of the car can be calculated using the formula, $a = (v\_f - v\_i) / t$, where v_i is the initial velocity, v_f is the final velocity (which is 0), and t is the time taken to come to a stop.

v_i = 20 m/s and t = 5 seconds.

a = (0 - 20) / 5 = -4 m/s^2.

Answer: the deceleration of the car is 4 m/s^2.

d) A stone is dropped from a height of 20 meters. How much time does it take to reach the ground?

The time taken by the stone to reach the ground can be calculated using the formula, t = sqrt(2 * h / g), where h is the initial height and g is the acceleration due to gravity.

h = 20 meters and g = 9.8 m/s^2. Substituting the values, we get t = sqrt(2 * 20 / 9.8) = 2 seconds.

Answer: it takes 2 seconds for the stone to reach the ground.

e) A box of mass 10 kg is pushed with a force of 50 N. What is the acceleration of the box?

The acceleration of the box can be calculated using the formula, $a = F\_net / m$, where F_net is the net force acting on the box and m is the mass of the box.

F_net = 50 N and m = 10 kg.

a = 50 N / 10 kg = 5 m/s^2.

Answer: the acceleration of the box is 5 m/s^2

Exercise 2

Questions

a) A cyclist travels a distance of 30 km in 1.5 hours. What is the average speed of the cyclist in kilometers per hour?

b) A ball is thrown horizontally with a speed of 10 m/s from a height of 5 meters. How far does the ball travel before it hits the ground?

c) What is the SI unit of force?

d) What is the weight of a body with a mass of 20 kg on the Earth's surface?

e) What is the difference between speed and velocity?

Answers

a) A cyclist travels a distance of 30 km in 1.5 hours. What is the average speed of the cyclist in kilometers per hour?

The average speed of the cyclist in kilometers per hour can be calculated by dividing the distance traveled by the time taken and converting the units. Answer: the average speed of the cyclist is (30 km / 1.5 hours) = 20 km/hr.

b) A ball is thrown horizontally with a speed of 10 m/s from a height of 5 meters. How far does the ball travel before it hits the ground?

The distance traveled by the ball before it hits the ground can be calculated using the formula, $d = v\_i * t$, where v_i is the initial horizontal velocity and t is the time taken to reach the ground.

v_i = 10 m/s and t can be calculated using the formula, $t = sqrt(2 * h / g)$, where h is the initial height and g is the acceleration due to gravity.

h = 5 meters and g = 9.8 m/s^2.

$t = sqrt(2 * 5 / 9.8)$ = 1 second.

Answer: $d = v\_i * t$ = 10 m/s * 1 second = 10 meters.

c) What is the SI unit of force?

Answer: the SI unit of force is Newton (N).

d) What is the weight of a body with a mass of 20 kg on the Earth's surface?

The weight of a body can be calculated using the formula, $W = m * g$, where m is the mass of the body and g is the acceleration due to gravity on the Earth's surface (9.8 m/s^2).

Answer: W = 20 kg * 9.8 m/s^2 = 196 N.

e) What is the difference between speed and velocity?

Speed is the rate at which an object travels, while velocity is the rate at which an object travels in a particular direction.

Exercise 3

Questions

a) What is the definition of work in physics?

b) What is the principle of conservation of energy?

c) What is the definition of acceleration in physics?

d) What is the SI unit of distance?

e) What is the formula for calculating the average speed of an object?

Answers

a) What is the definition of work in physics?

Answer: In physics, work is defined as the product of force and displacement in the direction of the force.

b) What is the principle of conservation of energy?

Answer: The principle of conservation of energy states that the total energy of a closed system remains constant, that is, energy cannot be created or destroyed but can only be transformed from one form to another.

c) What is the definition of acceleration in physics?

Answer: Acceleration is the rate at which the velocity of an object changes with respect to time.

d) What is the SI unit of distance?

Answer: The SI unit of distance is meter (m).

e) What is the formula for calculating the average speed of an object?

Answer: The formula for calculating the average speed of an object is : average speed = total distance traveled / total time taken.

Exercise 4

Questions

a) What is the difference between kinetic and potential energy?

b) What is the SI unit of work and energy?

c) A car travels a distance of 100 meters in 10 seconds. What is its average speed?

d) What is the SI unit of time?

e) What is the difference between mass and weight?

Answers

a) What is the difference between kinetic and potential energy?

Answer: Kinetic energy is the energy possessed by a moving object due to its motion, while potential energy is the energy possessed by an object due to its position or configuration.

b) What is the SI unit of work and energy?

Answer: The SI unit of work and energy is Joule (J).

c) A car travels a distance of 100 meters in 10 seconds. What is its average speed?

Answer: The average speed of the car can be calculated as:

average speed = total distance traveled / total time taken = 100 m / 10 s = 10 m/s.

d) What is the SI unit of time?

Answer: The SI unit of time is second (s)

e) What is the difference between mass and weight?

Answer: Mass is the amount of matter in an object, while weight is the force with which an object is attracted towards the Earth due to gravity.

Exercise 5

Questions

a) What is the formula for calculating the distance traveled by an object with constant acceleration?

b) What is the difference between static and dynamic friction?

c) A ball is thrown upwards with an initial velocity of 10 m/s. What is the maximum height it can reach and how long does it take to reach that height?

d) A car of mass 1000 kg is moving with a velocity of 20 m/s. If the brakes are applied, and the car comes to a stop in 5 seconds, what is the force applied by the brakes?

e) A 2 kg block is placed on a horizontal surface with a coefficient of static friction of 0.4. What is the maximum force that can be applied to the block to prevent it from sliding?

Answers

a) What is the formula for calculating the distance traveled by an object with constant acceleration?

Answer: the formula for calculating the distance traveled by an object with constant acceleration is given by:

distance = initial velocity * time + 1/2 * acceleration * time^2

b) What is the difference between static and dynamic friction?

Answer: static friction is the frictional force that opposes the motion of an object at rest, while dynamic friction is the frictional force that opposes the motion of an object in motion.

c) A ball is thrown upwards with an initial velocity of 10 m/s. What is the maximum height it can reach and how long does it take to reach that height?

Answer: Using the formula for maximum height, h = v^2 / 2g, where v is the initial velocity and g is the acceleration due to gravity (9.8 m/s^2), the maximum height the ball can reach is h = (10 m/s)^2 / 2 * 9.8 m/s^2 = 5.1 m.

Using the formula for time of flight, t = 2v/g, we can calculate that the time it takes for the ball to reach that height is t = 2 * 10 m/s / 9.8 m/s^2 = 2.0 s

Answer: 2.0 s

d) A car of mass 1000 kg is moving with a velocity of 20 m/s. If the brakes are applied, and the car comes to a stop in 5 seconds, what is the force applied by the brakes?

Using the formula for force, F = ma, where m is the mass of the car and a is the deceleration due to the brakes, we can calculate that the deceleration of the car is a = v/t = 20 m/s / 5 s = 4 m/s^2.

F = 1000 kg * 4 m/s^2 = 4000 N.

Answer: 4000 N

e) A 2 kg block is placed on a horizontal surface with a coefficient of static friction of 0.4. What is the maximum force that can be applied to the block to prevent it from sliding?

Using the formula for maximum force of static friction, Fmax = μs * N, where μs is the coefficient of static friction and N is the normal force, we can calculate that the maximum force that can be applied to the block to prevent it from sliding : Fmax = 0.4 * 2 kg * 9.8 m/s^2 = 7.84 N

Answer: 7.84 N

Exercise 1

Questions

a) A spring with a spring constant of 200 N/m is compressed by 0.1 m. What is the potential energy stored in the spring?

b) An object of mass 5 kg is moving with a velocity of 10 m/s. What is its kinetic energy?

c) A block of mass 2 kg is released from a height of 10 m above the ground. What is the speed of the block just before it hits the ground?

d) A bullet of mass 0.01 kg is fired from a gun with a velocity of 1000 m/s. If the gun has a mass of 1 kg, what is the recoil velocity of the gun?

e) A car of mass 1500 kg is moving at a velocity of 20 m/s. If the engine delivers a force of 5000 N, what is the acceleration of the car?

Answers

a) A spring with a spring constant of 200 N/m is compressed by 0.1 m. What is the potential energy stored in the spring?

Using the formula for potential energy stored in a spring, $U = 1/2 * k * x^2$, where k is the spring constant and x is the displacement, the potential energy stored in the spring is $U = 1/2 * 200$ N/m $* (0.1$ m$)^2 = 1$ J.

Answer: 1J

b) An object of mass 5 kg is moving with a velocity of 10 m/s. What is its kinetic energy?

Using the formula for kinetic energy, $K = 1/2 * mv^2$, where m is the mass of the object and v is its velocity, the kinetic energy of the object is:

$K = 1/2 * 5$ kg $* (10$ m/s$)^2 = 250$ J

Answer: 250 J

c) A block of mass 2 kg is released from a height of 10 m above the ground. What is the speed of the block just before it hits the ground?

Using the formula for potential energy, mgh, where m is the mass of the block, g is the acceleration due to gravity (9.8 m/s^2), and h is the height of the block above the ground, we can calculate that the potential energy of the block at the top is U = 2 kg * 9.8 m/s^2 * 10 m = 196 J.

Using the formula for kinetic energy, 1/2 * mv^2, we can equate the initial potential energy to the final kinetic energy just before the block hits the ground.

Answer: v = sqrt(2gh) = sqrt(2 * 9.8 m/s^2 * 10 m) = 14 m/s.

d) A bullet of mass 0.01 kg is fired from a gun with a velocity of 1000 m/s. If the gun has a mass of 1 kg, what is the recoil velocity of the gun?

Using the conservation of momentum, m1v1 + m2v2 = 0, where m1 and v1 are the mass and velocity of the bullet, and m2 and v2 are the mass and velocity of the gun, the recoil velocity of the gun is:

Answer: v2 = -m1v1/m2 = -(0.01 kg)(1000 m/s)/(1 kg) = -10 m/s (negative sign indicating the direction opposite to that of the bullet).

e) A car of mass 1500 kg is moving at a velocity of 20 m/s. If the engine delivers a force of 5000 N, what is the acceleration of the car?

Using the formula for force, F = ma, where F is the force applied, m is the mass of the car, and a is the acceleration, the acceleration of the car is a = F/m = 5000 N / 1500 kg = 3.33 m/s^2

Answer: 3.33 m/s^2

Exercise 2

Questions

a) A ball of mass 0.1 kg is attached to a string and swung in a circle of radius 0.5 m with a velocity of 5 m/s. What is the tension in the string?

b) A block of mass 3 kg is pulled up a 30° incline with a force of 20 N. If the coefficient of kinetic friction is 0.2, what is the acceleration of the block?

c) A block of mass 4 kg is pushed along a horizontal surface with a force of 10 N. If the coefficient of kinetic friction is 0.3, what is the acceleration of the block?

d) A car of mass 1000 kg is moving at a velocity of 30 m/s. If the engine delivers a power of 50 kW, what is the acceleration of the car?

e) A block of mass 2 kg is dropped from a height of 5 m above a spring of spring constant 2000 N/m. What is the maximum compression of the spring when the block hits it?

Answers

a) A ball of mass 0.1 kg is attached to a string and swung in a circle of radius 0.5 m with a velocity of 5 m/s. What is the tension in the string?

Using the formula for centripetal force, $Fc = mv^2/r$, where m is the mass of the ball, v is its velocity, and r is the radius of the circle, the centripetal force acting on the ball is $Fc = (0.1 \text{ kg})(5 \text{ m/s})^2 / 0.5 \text{ m} = 5 \text{ N}$.

Answer: since the string is the only force acting on the ball in the radial direction, the tension in the string is equal to the centripetal force, so the tension in the string is 5 N.

b) A block of mass 3 kg is pulled up a 30° incline with a force of 20 N. If the coefficient of kinetic friction is 0.2, what is the acceleration of the block?

Using the formula for force along the incline, $Fpar = mg \sin(\theta)$, where m is the mass of the block, g is the acceleration due to gravity, and θ is the angle of the incline, we can calculate that the force along the incline is:

Answer: $Fpar = (3 \text{ kg})(9.8 \text{ m/s}^2) \sin(30°) = 14.7 \text{ N}$

c) A block of mass 4 kg is pushed along a horizontal surface with a force of 10 N. If the coefficient of kinetic friction is 0.3, what is the acceleration of the block?

Using the formula for force along the horizontal direction, $Fpar = ma$, where Fpar is the parallel force applied, m is the mass of the block, and a is the acceleration, the parallel force applied is $Fpar = 10 \text{ N}$.

The force of friction is $Ffric = \mu k * Fn$, where μk is the coefficient of kinetic friction and Fn is the normal force acting on the block, which is equal to the weight of the block (mg).

Thus, $Ffric = (0.3)(4 \text{ kg})(9.8 \text{ m/s}^2) = 11.76 \text{ N}$.

The net force acting on the block is $Fnet = Fpar - Ffric = 10 \text{ N} - 11.76 \text{ N} = -1.76$ N (negative sign indicating opposite direction to motion).

Answer: the acceleration of the block is $a = Fnet/m = (-1.76 \text{ N})/(4 \text{ kg}) = -0.44$ m/s^2.

d) A car of mass 1000 kg is moving at a velocity of 30 m/s. If the engine delivers a power of 50 kW, what is the acceleration of the car?

Using the formula for power, $P = Fv$, where P is the power delivered, F is the force applied, and v is the velocity, we can calculate that the force applied is $F = P/v = 50,000$ W / 30 m/s = 1666.67 N. Using the formula for force, $F = ma$, Answer: the acceleration of the car is a = F/m = 1666.67 N / 1000 kg = 1.67 m/s^2.

e) A block of mass 2 kg is dropped from a height of 5 m above a spring of spring constant 2000 N/m. What is the maximum compression of the spring when the block hits it?

Using the formula for potential energy, mgh, where m is the mass of the block, g is the acceleration due to gravity (9.8 m/s^2), and h is the height of the block above the spring, the potential energy of the block at the top is U = 2 kg * 9.8 m/s^2 * 5 m = 98 J.

The maximum compression of the spring occurs when all of the potential energy of the block is converted to elastic potential energy of the spring.

Using the formula for elastic potential energy, 1/2 kx^2, where k is the spring constant and x is the maximum compression of the spring, we can equate the initial potential energy of the block to the final elastic potential energy of the spring.

Answer: x = sqrt(2U/k) = sqrt(2*98 J / 2000 N/m) = 0.7 m.

Exercise 3

Questions

a) A ball of mass 0.2 kg is dropped from a height of 1 m above a horizontal surface. If the coefficient of restitution between the ball and the surface is 0.6, what is the velocity of the ball just before it hits the surface the second time?

b) A pendulum of length 1 m and mass 0.1 kg swings back and forth with a period of 2 seconds. What is the tension in the string when the pendulum is at its lowest point?

c) A car travels at a constant speed of 50 km/h for 2 hours. How far does it travel?

d) A ball is thrown straight up into the air with an initial velocity of 20 m/s. How high does it go?

e) A book weighing 5 N is placed on a table. What is the normal force exerted by the table on the book?

Answers

a) A ball of mass 0.2 kg is dropped from a height of 1 m above a horizontal surface. If the coefficient of restitution between the ball and the surface is 0.6, what is the velocity of the ball just before it hits the surface the second time?

Using the formula for potential energy, mgh, where m is the mass of the ball, g is the acceleration due to gravity, and h is the height of the ball above the surface, we can calculate that the potential energy of the ball at the top is :

$U = 0.2 \text{ kg} * 9.8 \text{ m/s}^2 * 1 \text{ m} = 1.96 \text{ J}$.

The velocity of the ball just before it hits the surface the first time is given by the equation for the conservation of mechanical energy, which states that the initial potential energy is equal to the final kinetic energy plus the final potential energy.

Thus, $1/2 \, mv1^2 = mgh - U$, where v1 is the velocity just before the ball hits the surface the first time.

Solving for v1, we get $v1 = \text{sqrt}(2gh - 2U/m) = \text{sqrt}(29.8 \text{ m/s}^2 * 1 \text{ m} - 21.96 \text{ J} / 0.2 \text{ kg}) = 3.92 \text{ m/s}$. Using the coefficient of restitution, $e = v2/v1$, where v2 is the velocity just after the ball hits the surface the first time, we can calculate that the velocity just after the ball hits the surface the first time is:

$v2 = ev1 = 0.6 * 3.92 \text{ m/s} = 2.35 \text{ m/s}$.

Answer: the velocity of the ball just before it hits the surface the second time is equal to the velocity just after it bounces off the surface the first time, so it is also 2.35 m/s.

b) A pendulum of length 1 m and mass 0.1 kg swings back and forth with a period of 2 seconds. What is the tension in the string when the pendulum is at its lowest point?

The period of a pendulum is given by the formula $T = 2\pi \text{sqrt}(L/g)$, where L is the length of the pendulum and g is the acceleration due to gravity.

Solving for g, we get $g = 4\pi^2 \, L/T^2 = 4\pi^2 * 1 \text{ m} / (2 \text{ s})^2 = 9.87 \text{ m/s}^2$.

At the lowest point of the pendulum's swing, the tension in the string is equal to the weight of the mass:

Answer: T = mg = 0.1 kg * 9.87 m/s^2 = 0.987 N

c) A car travels at a constant speed of 50 km/h for 2 hours. How far does it travel?

Distance = Speed x Time.

Answer: Distance = 50 km/h x 2 h = 100 km.

d) A ball is thrown straight up into the air with an initial velocity of 20 m/s. How high does it go?

The maximum height is reached when the velocity is zero.

The formula for maximum height is h = (v^2) / (2g), where v is the initial velocity and g is the acceleration due to gravity.

Answer: h = (20 m/s)^2 / (2 x 9.8 m/s^2) = 20.41 m

e) A book weighing 5 N is placed on a table. What is the normal force exerted by the table on the book?

Answer: the normal force is equal and opposite to the weight of the book, which is 5 N

Exercise 4

Questions

a) A spring with a spring constant of 10 N/m is compressed by 2 cm. What is the potential energy stored in the spring?

b) A box weighing 100 N is pushed with a force of 50 N at an angle of 30 degrees to the horizontal. What is the net force on the box?

c) A ball is thrown horizontally from the top of a cliff with an initial velocity of 15 m/s. The cliff is 25 m high. How far from the base of the cliff does the ball land?

d) A spring with a spring constant of 100 N/m is stretched by 5 cm. What is the force exerted by the spring?

e) A car of mass 1000 kg is traveling at a speed of 20 m/s. If the brakes are applied and the car comes to a stop in 5 seconds, what is the average braking force?

Answers

a) A spring with a spring constant of 10 N/m is compressed by 2 cm. What is the potential energy stored in the spring?

The potential energy stored in a spring is given by the formula $U = (1/2) k x^2$, where k is the spring constant and x is the displacement from the equilibrium position.

Answer: $U = (1/2) \times k x^2 = (1/2) \times 10$ N/m $\times (0.02$ m$)^2 = 0.002$ J.

b) A box weighing 100 N is pushed with a force of 50 N at an angle of 30 degrees to the horizontal. What is the net force on the box?

The horizontal component of the force is $Fx = F \cos(theta) = 50$ N $\times \cos(30$ degrees$) = 43.3$ N.

The vertical component of the force is $Fy = F \sin(theta) = 50$ N $\times \sin(30$ degrees$) = 25$ N.

The weight of the box is 100 N downwards.

Answer: the net force is the vector sum of the horizontal and vertical forces, which is $sqrt(Fx^2 + Fy^2 - W^2) = sqrt((43.3$ N$)^2 + (25$ N$)^2 - (100$ N$)^2) = 35.3$ N

c) A ball is thrown horizontally from the top of a cliff with an initial velocity of 15 m/s. The cliff is 25 m high. How far from the base of the cliff does the ball land?

The time it takes for the ball to fall to the ground is given by $t = sqrt(2h/g)$, where h is the height of the cliff and g is the acceleration due to gravity.

$t = sqrt(2 \times 25$ m $/ 9.8$ m/s$^2) = 2.26$ s.

During this time, the ball travels a horizontal distance of $d = vt$,

where v is the initial horizontal velocity.

Answer: $d = 15$ m/s $\times 2.26$ s $= 33.9$ m.

d) A spring with a spring constant of 100 N/m is stretched by 5 cm. What is the force exerted by the spring?

The force exerted by a spring is given by $F = kx$, where k is the spring constant and x is the displacement from the equilibrium position

Answer: $F = 100$ N/m $\times 0.05$ m $= 5$ N

e) A car of mass 1000 kg is traveling at a speed of 20 m/s. If the brakes are applied and the car comes to a stop in 5 seconds, what is the average braking force?

The initial kinetic energy of the car is given by $KE = (1/2) mv^2$, where m is the mass of the car and v is the velocity.

$KE = (1/2) \times 1000$ kg $\times (20$ m/s$)^2 = 200,000$ J.

The final kinetic energy of the car is zero, since it comes to a stop.

The work done by the braking force is equal to the initial kinetic energy.

The work done by a constant force is given by W = Fd, where d is the distance over which the force acts.

The distance is given by d = (1/2) at^2, where a is the acceleration and t is the time.

d = (1/2) x (-4 m/s^2) x (5 s)^2 = -50 m (note that the negative sign is because the car is slowing down).

W = Fd = -200,000 J.

Answer: F = -200,000 J / -50 m = 4000 N.

The negative sign indicates that the force is in the opposite direction to the motion of the car.

Exercise 5

Questions

a) A block of mass 2 kg is pushed with a force of 20 N on a horizontal surface with a coefficient of friction of 0.3. What is the acceleration of the block?

b) A ball is thrown vertically upwards with an initial velocity of 20 m/s. What is the maximum height reached by the ball?

c) A car travels a distance of 50 m in 10 seconds. What is its average speed?

d) A ball is dropped from a height of 10 m. What is its speed just before it hits the ground?

e) A block of mass 2 kg is pushed with a force of 10 N on a horizontal surface with no friction. What is the acceleration of the block?

Answers

a) A block of mass 2 kg is pushed with a force of 20 N on a horizontal surface with a coefficient of friction of 0.3. What is the acceleration of the block?

The force of friction is given by f = uN, where u is the coefficient of friction and N is the normal force.

The normal force is equal to the weight of the block, which is N = mg = 2 kg x 9.8 m/s^2 = 19.6 N.

f = 0.3 x 19.6 N = 5.88 N.

The net force on the block is F = 20 N - 5.88 N = 14.12 N.

Answer: the acceleration of the block is = F/m = 14.12 N / 2 kg = 7.06 m/s^2.

b) A ball is thrown vertically upwards with an initial velocity of 20 m/s. What is the maximum height reached by the ball?

The maximum height reached by the ball is given by h = (v^2)/(2g), where v is the initial velocity and g is the acceleration due to gravity.

Answer: h = (20 m/s)^2 / (2 x 9.8 m/s^2) = 20.4 m

c) A car travels a distance of 50 m in 10 seconds. What is its average speed?

The average speed of the car is given by v = d/t, where d is the distance traveled and t is the time taken.

Answer: v = 50 m / 10 s = 5 m/s

d) A ball is dropped from a height of 10 m. What is its speed just before it hits the ground?

The speed of the ball just before it hits the ground is given by v = sqrt(2gh), where g is the acceleration due to gravity and h is the height from which the ball is dropped.

Answer: v = sqrt(2 x 9.8 m/s^2 x 10 m) = 14 m/s.

e) A block of mass 2 kg is pushed with a force of 10 N on a horizontal surface with no friction. What is the acceleration of the block?

The net force on the block is given by F = ma, where m is the mass of the block and a is the acceleration.

10 N = 2 kg x a

Answer: a = 5 m/s^2

Exercise 1

Questions

a) A stone is thrown horizontally with an initial velocity of 10 m/s from a height of 5 m. How far from the base of the building does the stone land?

b) A car travels a distance of 20 meters in 4 seconds. What is its average speed?

c) A ball is dropped from a height of 2 meters. What is its speed just before it hits the ground?

d) A block of mass 1 kg is pushed with a force of 5 N on a horizontal surface with no friction. What is the acceleration of the block?

e) A train moves at a speed of 10 m/s for 30 seconds. How far does it travel in this time?

Answers

a) A stone is thrown horizontally with an initial velocity of 10 m/s from a height of 5 m. How far from the base of the building does the stone land?

The time taken for the stone to fall is given by $t = \text{sqrt}(2h/g)$, where h is the height from which the stone is thrown and g is the acceleration due to gravity.

$t = \text{sqrt}(2 \times 5 \text{ m} / 9.8 \text{ m/s}^2) = 1.01$ s.

During this time, the stone travels a horizontal distance of $d = vt$, where v is the initial horizontal velocity.

Answer: $d = 10 \text{ m/s} \times 1.01 \text{ s} = 10.1$ m

b) A car travels a distance of 20 meters in 4 seconds. What is its average speed?

The average speed of the car is given by $v = d/t$, where d is the distance traveled and t is the time taken.

$v = 20 \text{ m} / 4 \text{ s} = 5$ m/s

c) A ball is dropped from a height of 2 meters. What is its speed just before it hits the ground?

The speed of the ball just before it hits the ground is given by $v = \sqrt{2gh}$, where g is the acceleration due to gravity and h is the height from which the ball is dropped.

$v = \sqrt{2 \times 9.8 \text{ m/s}^2 \times 2 \text{ m}} = 6.26$ m/s.

d) A block of mass 1 kg is pushed with a force of 5 N on a horizontal surface with no friction. What is the acceleration of the block?

The net force on the block is given by $F = ma$, where m is the mass of the block and a is the acceleration.

5 N = 1 kg x a

Answer: $a = 5$ m/s^2

e) A train moves at a speed of 10 m/s for 30 seconds. How far does it travel in this time?

The distance traveled by the train is given by $d = vt$, where v is the speed and t is the time taken.

d = 10 m/s x 30 s = 300 m.

Exercise 2

Questions

a) A stone is thrown vertically upwards with an initial velocity of 5 m/s. What is the maximum height reached by the stone?

b) A ball is thrown horizontally from the top of a building with a speed of 10 m/s. The building is 20 meters tall. How far from the base of the building does the ball land?

c) A block of mass 2 kg is suspended by a rope. A force of 20 N is applied to the block at an angle of 30 degrees with the vertical. What is the tension in the rope?

d) A car of mass 1000 kg is traveling at a speed of 20 m/s. What is the kinetic energy of the car?

e) A spring with a spring constant of 100 N/m is compressed by 0.1 meters. What is the potential energy stored in the spring?

Answers

a) A stone is thrown vertically upwards with an initial velocity of 5 m/s. What is the maximum height reached by the stone?

The maximum height reached by the stone is given by $h = (v^2)/(2g)$, where v is the initial velocity and g is the acceleration due to gravity.

Answer: $h = (5 \text{ m/s})^2 / (2 \times 9.8 \text{ m/s}^2) = 1.27$ m

b) A ball is thrown horizontally from the top of a building with a speed of 10 m/s. The building is 20 meters tall. How far from the base of the building does the ball land?

The time taken for the ball to hit the ground is given by $t = \text{sqrt}(2h/g)$, where h is the height of the building and g is the acceleration due to gravity.

$t = \text{sqrt}(2 \times 20 \text{ m} / 9.8 \text{ m/s}^2) = 2.02$ s.

The horizontal distance traveled by the ball is given by $d = vt$, where v is the horizontal velocity of the ball.

Since the ball is thrown horizontally, the horizontal velocity is constant and equal to 10 m/s.

$d = 10 \text{ m/s} \times 2.02 \text{ s} = 20.2$ m

Answer: the ball lands 20.2 meters from the base of the building

c) A block of mass 2 kg is suspended by a rope. A force of 20 N is applied to the block at an angle of 30 degrees with the vertical. What is the tension in the rope?

The tension in the rope is given by $T = mg + F\sin(\text{theta})$, where m is the mass of the block, g is the acceleration due to gravity, F is the applied force, and theta is the angle between the force and the vertical.

$T = 2 \text{ kg} \times 9.8 \text{ m/s}^2 + 20 \text{ N} \times \sin(30 \text{ degrees}) = 19.6 \text{ N} + 10 \text{ N} = 29.6$ N.

Answer: the tension in the rope is 29.6 N.

d) A car of mass 1000 kg is traveling at a speed of 20 m/s. What is the kinetic energy of the car?

The kinetic energy of the car is given by $KE = (1/2)mv^2$, where m is the mass of the car and v is its velocity.

$KE = (1/2) \times 1000 \text{ kg} \times (20 \text{ m/s})^2 = 200{,}000$ J.

Answer: the kinetic energy of the car is 200,000 Joules.

e) A spring with a spring constant of 100 N/m is compressed by 0.1 meters. What is the potential energy stored in the spring?

The potential energy stored in a spring is given by PE = (1/2)kx^2, where k is the spring constant and x is the displacement of the spring from its equilibrium position.

PE = (1/2) x 100 N/m x (0.1 m)^2 = 0.5 J.

Answer: the potential energy stored in the spring is 0.5 Joules.

Exercise 3

Questions

a) A rocket of mass 1000 kg is launched vertically upwards with a force of 10,000 N. What is the acceleration of the rocket?

b) A car travels 100 km at a constant speed of 50 km/h. What is its average velocity?

c) An object moves 20 m in 5 seconds with a constant velocity. What is its speed?

d) A ball is thrown upward with a velocity of 20 m/s. How long will it take to reach its maximum height?

e) A stone is thrown horizontally from the top of a 20 m high cliff with a velocity of 10 m/s. How far from the base of the cliff will it land?

f) A car is moving at a constant velocity of 20 m/s for 10 seconds. What is its displacement during this time?

Answers

a) A rocket of mass 1000 kg is launched vertically upwards with a force of 10,000 N. What is the acceleration of the rocket?

The net force on the rocket is given by F = ma, where m is the mass of the rocket and a is its acceleration.

10,000 N - 1000 kg x 9.8 m/s^2 = 9020 N.

Solving for a, we get a = F/m = 9020 N / 1000 kg = 9.02 m/s^2.

Answer: the acceleration of the rocket is 9.02 m/s^2.

b) A car travels 100 km at a constant speed of 50 km/h. What is its average velocity?

Answer: Average velocity = Total displacement/Total time = 100 km/2 h = 50 km/h

c) An object moves 20 m in 5 seconds with a constant velocity. What is its speed?

Answer: Speed = distance/time = 20 m/5 s = 4 m/s

d) A ball is thrown upward with a velocity of 20 m/s. How long will it take to reach its maximum height?

The time it takes to reach maximum height is given by: $t = v/g$, where v is the initial velocity and g is the acceleration due to gravity.

Answer: $t = 20$ m/s / 9.8 m/s^2 = 2.04 seconds.

e) A stone is thrown horizontally from the top of a 20 m high cliff with a velocity of 10 m/s. How far from the base of the cliff will it land?

The time taken for the stone to reach the ground is given by: $t = \sqrt{2h/g}$, where h is the height of the cliff and g is the acceleration due to gravity.

$t = \sqrt{2 \times 20 \text{ m}/9.8 \text{ m/s}^2} = 2.02$ seconds.

In this time, the stone would travel a horizontal distance of $d = vt$, where v is the horizontal velocity, which is 10 m/s in this case.

Answer: $d = 10$ m/s x 2.02 s = 20.2 meters.

f) A car is moving at a constant velocity of 20 m/s for 10 seconds. What is its displacement during this time?

Answer: Displacement = velocity x time = 20 m/s x 10 s = 200 meters.

Exercise 4

Questions

a) An object falls freely from a height of 10 meters. What is its velocity when it hits the ground?

b) A car accelerates from rest to a speed of 20 m/s in 5 seconds. What is its acceleration?

c) A person throws a ball vertically upward with a velocity of 10 m/s. What is the maximum height reached by the ball?

d) A car travels 50 kilometers in 2 hours at a constant speed. What is its average speed?

e) An object is moving with a velocity of 5 m/s. If its acceleration is 2 m/s^2, what is its velocity after 3 seconds?

Answers

a) An object falls freely from a height of 10 meters. What is its velocity when it hits the ground?

The final velocity of the object can be calculated using the equation v^2 = u^2 + 2as, where u is the initial velocity (which is zero since it was dropped), s is the distance fallen (10 meters), and a is the acceleration due to gravity (-9.8 m/s^2).

v^2 = 0 + 2(-9.8 m/s^2)(10 m) = -196 m^2/s^2.

Answer: since velocity cannot be negative, we take the square root of the positive value of v^2, which gives v = 14 m/s.

b) A car accelerates from rest to a speed of 20 m/s in 5 seconds. What is its acceleration?

The acceleration of the car can be calculated using the equation a = (v - u)/t, where v is the final velocity (20 m/s), u is the initial velocity (0 m/s), and t is the time taken (5 seconds).

Answer: a = (20 m/s - 0 m/s)/5 s = 4 m/s^2

c) A person throws a ball vertically upward with a velocity of 10 m/s. What is the maximum height reached by the ball?

The maximum height reached by the ball can be calculated using the equation h = u^2/(2g), where u is the initial velocity (10 m/s) and g is the acceleration due to gravity (-9.8 m/s^2).

Answer: h = (10 m/s)^2/(2 x -9.8 m/s^2) = 5.1 meters.

d) A car travels 50 kilometers in 2 hours at a constant speed. What is its average speed?

The average speed of the car can be calculated using the equation average speed = total distance / total time.

Answer: the average speed = 50 km / 2 hours = 25 km/h.

e) An object is moving with a velocity of 5 m/s. If its acceleration is 2 m/s^2, what is its velocity after 3 seconds?

The final velocity of the object can be calculated using the equation v = u + at, where u is the initial velocity (5 m/s), a is the acceleration (2 m/s^2), and t is the time taken (3 seconds).

Answer: v = 5 m/s + 2 m/s^2 x 3 s = 11 m/s.

Exercise 5

Questions

a) A ball is dropped from a height of 5 meters. How long does it take to hit the ground?

b) A car travels a distance of 100 meters in 10 seconds. What is its speed?

c) An object is moving with a velocity of 2 m/s. If its acceleration is 3 m/s^2, what is its velocity after 4 seconds?

d) A person pushes a box with a force of 20 N on a flat surface. If the box has a mass of 5 kg, what is its acceleration?

e) A ball is thrown horizontally with a velocity of 10 m/s. How far does it travel in 2 seconds?

Answers

a) A ball is dropped from a height of 5 meters. How long does it take to hit the ground?

The time taken for the ball to hit the ground can be calculated using the equation t = sqrt(2h/g), where h is the height (5 meters) and g is the acceleration due to gravity (9.8 m/s^2).

Answer: t = sqrt(2 x 5 m / 9.8 m/s^2) = 1 second.

b) A car travels a distance of 100 meters in 10 seconds. What is its speed?

The speed of the car can be calculated using the equation speed = distance / time.

Answer: speed = 100 m / 10 s = 10 m/s.

c) An object is moving with a velocity of 2 m/s. If its acceleration is 3 m/s^2, what is its velocity after 4 seconds?

The final velocity of the object can be calculated using the equation v = u + at, where u is the initial velocity (2 m/s), a is the acceleration (3 m/s^2), and t is the time taken (4 seconds).

Answer: v = 2 m/s + 3 m/s^2 x 4 s = 14 m/s.

d) A person pushes a box with a force of 20 N on a flat surface. If the box has a mass of 5 kg, what is its acceleration?

The acceleration of the box can be calculated using the equation a = F/m, where F is the force (20 N) and m is the mass (5 kg).

Answer: a = 20 N / 5 kg = 4 m/s^2.

e) A ball is thrown horizontally with a velocity of 10 m/s. How far does it travel in 2 seconds?

The horizontal distance travelled by the ball can be calculated using the equation distance = velocity x time.

Since the velocity is horizontal, the distance travelled is simply the product of the velocity and time.

Answer: distance = 10 m/s x 2 s = 20 meters.

Exercise 1

Questions

a) What is the weight of a 10 kg object on Earth?

b) A ball is rolling down a hill with an acceleration of 2 m/s^2. If it starts from rest, what is its velocity after 5 seconds?

c) A person lifts a 20 kg box to a height of 2 meters. How much work is done?

d) An object is thrown vertically upwards with a velocity of 10 m/s. What is its maximum height?

e) A car accelerates from rest to a speed of 20 m/s in 5 seconds. What is its acceleration?

Answers

a) What is the weight of a 10 kg object on Earth?

The weight of an object on Earth can be calculated using the equation:

weight = mass x acceleration due to gravity.

The acceleration due to gravity on Earth is approximately 9.8 m/s^2.

Answer: the weight of a 10 kg object on Earth is 10 kg x 9.8 m/s^2 = 98 N.

b) A ball is rolling down a hill with an acceleration of 2 m/s^2. If it starts from rest, what is its velocity after 5 seconds?

The final velocity of the ball can be calculated using the equation v = u + at, where u is the initial velocity (0 m/s), a is the acceleration (2 m/s^2), and t is the time taken (5 seconds).

Answer: v = 0 m/s + 2 m/s^2 x 5 s = 10 m/s.

c) A person lifts a 20 kg box to a height of 2 meters. How much work is done?

The work done can be calculated using the equation work = force x distance x cos(theta), where force is the weight of the box (mass x acceleration due to gravity), distance is the height lifted, and theta is the angle between the force and the displacement (which is 0 degrees since the force and displacement are in the same direction).

Answer: the work done is (20 kg x 9.8 m/s^2) x 2 m x cos(0) = 392 J.

d) An object is thrown vertically upwards with a velocity of 10 m/s. What is its maximum height?

The maximum height reached by the object can be calculated using the equation h = (v^2 - u^2) / 2g, where v is the final velocity (0 m/s at maximum height), u is the initial velocity (10 m/s upwards), and g is the acceleration due to gravity (-9.8 m/s^2).

Answer: h = (0^2 - 10^2) / (2 x -9.8 m/s^2) = 5.1 meters.

e) A car accelerates from rest to a speed of 20 m/s in 5 seconds. What is its acceleration?

The acceleration of the car can be calculated using the equation a = (v - u) / t, where v is the final velocity (20 m/s), u is the initial velocity (0 m/s), and t is the time taken (5 seconds).

Answer: a = (20 m/s - 0 m/s) / 5 s = 4 m/s^2.

Exercise 2

Questions

a) A force of 20 N is applied to an object of mass 5 kg. What is its acceleration?

b) A force of 10 N is applied to an object of mass 2 kg, which is initially at rest. How much work is done after the object has moved a distance of 3 meters?

c) An object is thrown horizontally with a velocity of 20 m/s. What is the maximum height it can reach?

Answers

a) A force of 20 N is applied to an object of mass 5 kg. What is its acceleration?

The acceleration of the object can be calculated using the equation F = ma, where F is the force applied, m is the mass of the object, and a is the resulting acceleration.

Answer: a = F/m = 20 N / 5 kg = 4 m/s^2

b) A force of 10 N is applied to an object of mass 2 kg, which is initially at rest. How much work is done after the object has moved a distance of 3 meters?

The work done by the force can be calculated using the equation:

work = force x distance x cos(theta), where force is the applied force, distance is the distance moved, and theta is the angle between the force and the displacement (which is 0 degrees since the force and displacement are in the same direction).

Answer: the work done is 10 N x 3 m x cos(0) = 30 J.

c) An object is thrown horizontally with a velocity of 20 m/s. What is the maximum height it can reach?

The maximum height reached by the object can be calculated using the equation h = (v^2 x sin^2(theta)) / 2g, where v is the initial velocity, theta is the angle of projection (which is 0 degrees since the object is thrown horizontally), and g is the acceleration due to gravity (-9.8 m/s^2).

Answer: h = (20 m/s)^2 x sin^2(0) / (2 x -9.8 m/s^2) = 0 meters.

Exercise 3

Questions

a) A ball is thrown straight up in the air. Neglecting air resistance, what is the acceleration of the ball at the highest point of its trajectory?

b) A car is traveling at a constant speed of 50 km/h. What is the car's acceleration?

c) A rock is dropped from the top of a cliff. Neglecting air resistance, what is the acceleration of the rock just before it hits the ground?

d) A force of 10 N is applied to an object with a mass of 2 kg. What is the object's acceleration?

e) A person is pushing a box with a force of 20 N, but the box is not moving. What is the frictional force acting on the box?

Answers

a) A ball is thrown straight up in the air. Neglecting air resistance, what is the acceleration of the ball at the highest point of its trajectory?

At the highest point of the trajectory, the ball momentarily comes to rest before falling back down.

its acceleration is equal to the acceleration due to gravity

Answer: approximately 9.8 m/s^2.

b) A car is traveling at a constant speed of 50 km/h. What is the car's acceleration?

Answer: the car's acceleration is zero, since it is traveling at a constant speed and not changing its velocity.

c) A rock is dropped from the top of a cliff. Neglecting air resistance, what is the acceleration of the rock just before it hits the ground?

Answer: the acceleration of the rock just before it hits the ground is approximately 9.8 m/s^2, which is the acceleration due to gravity.

d) A force of 10 N is applied to an object with a mass of 2 kg. What is the object's acceleration?

The object's acceleration can be calculated using the formula a = F/m, where F is the force applied and m is the mass of the object.

Answer: a = 10 N / 2 kg = 5 m/s^2.

e) A person is pushing a box with a force of 20 N, but the box is not moving. What is the frictional force acting on the box?

The frictional force acting on the box is equal and opposite to the pushing force, which is 20 N.

Answer: the frictional force is also 20 N

Exercise 4

Questions

a) A car travels 60 km in one hour. What is its average speed?

b) An object moves with a constant velocity of 5 m/s. What is its acceleration?

c) A ball is thrown horizontally from a height of 2 meters with an initial velocity of 10 m/s. Neglecting air resistance, how long does it take for the ball to hit the ground?

d) An object is moving with a velocity of 10 m/s and then comes to a stop in 5 seconds. What is its acceleration?

e) A force of 20 N is applied to an object with a mass of 5 kg. What is the object's acceleration?

Answers

a) A car travels 60 km in one hour. What is its average speed?

Average speed = total distance / total time = 60 km / 1 hour = 60 km/h.

Answer: 60 km/h.

b) An object moves with a constant velocity of 5 m/s. What is its acceleration?

The object's acceleration is zero, since it is moving with a constant velocity and not changing its speed or direction.

c) A ball is thrown horizontally from a height of 2 meters with an initial velocity of 10 m/s. Neglecting air resistance, how long does it take for the ball to hit the ground?

The time it takes for the ball to hit the ground can be found using the formula:

$t = sqrt(2h/g)$, where h is the initial height and g is the acceleration due to gravity.

Answer: $t = sqrt(2*2/9.8) = 0.64$ seconds.

d) An object is moving with a velocity of 10 m/s and then comes to a stop in 5 seconds. What is its acceleration?

The object's acceleration can be calculated using the formula:

$a = (v\_f - v\_i) / t$, where v_f is the final velocity, v_i is the initial velocity, and t is the time.

Answer: $a = (0 - 10)$ m/s / 5 s = -2 m/s^2

e) A force of 20 N is applied to an object with a mass of 5 kg. What is the object's acceleration?

The object's acceleration can be calculated using the formula $a = F/m$, where F is the force applied and m is the mass of the object.

Answer: $a = 20$ N / 5 kg = 4 m/s^2

Exercise 5

Questions

a) A car accelerates from rest to 20 m/s in 5 seconds. What is its acceleration?

b) An object is dropped from a height of 10 meters. How long does it take to hit the ground? (Neglect air resistance.)

c) A ball is thrown straight up with a velocity of 20 m/s. What is the maximum height the ball reaches? (Neglect air resistance.)

d) A force of 10 N is applied to an object with a mass of 2 kg. What is the object's acceleration?

e) A truck travels 500 km in 10 hours. What is its average speed?

Answers

a) A car accelerates from rest to 20 m/s in 5 seconds. What is its acceleration?

The car's acceleration can be calculated using the formula $a = (v\_f - v\_i) / t$, where v_f is the final velocity, v_i is the initial velocity, and t is the time.

Answer: $a = (20 \text{ m/s} - 0 \text{ m/s}) / 5 \text{ s} = 4 \text{ m/s}^2$.

b) An object is dropped from a height of 10 meters. How long does it take to hit the ground? (Neglect air resistance.)

The time it takes for the object to hit the ground can be found using the formula $t = sqrt(2h/g)$, where h is the initial height and g is the acceleration due to gravity.

Answer: $t = sqrt(2*10/9.8) = 1.43$ seconds.

c) A ball is thrown straight up with a velocity of 20 m/s. What is the maximum height the ball reaches? (Neglect air resistance.)

The maximum height the ball reaches can be found using the formula:

$h\_max = (v\_i^2) / (2g)$, where v_i is the initial velocity and g is the acceleration due to gravity.

Answer: $h\_max = (20 \text{ m/s})^2 / (2*9.8 \text{ m/s}^2) = 20.4$ meters.

d) A force of 10 N is applied to an object with a mass of 2 kg. What is the object's acceleration?

The object's acceleration can be calculated using the formula $a = F/m$, where F is the force applied and m is the mass of the object.

Answer: $a = 10 \text{ N} / 2 \text{ kg} = 5 \text{ m/s}^2$

e) A truck travels 500 km in 10 hours. What is its average speed?

Average speed = total distance / total time = 500 km / 10 hours = 50 km/h.

Answer: 50 km/h.

Exercise 1

Questions

a) An object is moving at a constant velocity of 10 m/s. How far does it travel in 20 seconds?

b) A ball is thrown horizontally with an initial velocity of 15 m/s from a height of 5 meters. Neglecting air resistance, how far does the ball travel before hitting the ground?

Answers

a) An object is moving at a constant velocity of 10 m/s. How far does it travel in 20 seconds?

The distance the object travels can be calculated using the formula

$d = v * t$, where v is the velocity and t is the time.

$d = 10$ m/s $* 20$ s $= 200$ meters.

b) A ball is thrown horizontally with an initial velocity of 15 m/s from a height of 5 meters. Neglecting air resistance, how far does the ball travel before hitting the ground?

The time it takes for the ball to hit the ground can be found using the formula:

$t = sqrt(2h/g)$, where h is the initial height and g is the acceleration due to gravity.

$t = sqrt(2*5/9.8) = 1.01$ seconds.

The horizontal distance the ball travels can be calculated using the formula:

$d = v * t$, where v is the initial horizontal velocity and t is the time.

Answer: $d = 15$ m/s $* 1.01$ s $= 15.15$ meters.

Conclusion

Thank you once again for purchasing this book. I hope it has helped you in your journey to understand the basics of mechanics.

Please, if you learnt something from this book, I would like you to leave a review. It'd be appreciated.

Thank you.

www.ingramcontent.com/pod-product-compliance
Lightning Source LLC
Chambersburg PA
CBHW080907220526
45466CB00011BA/3498